Tucholsky Wagner   Zola   Scott   Sydow   Schlegel
Turgenev   Wallace   Fonatne   Freud
        Twain   Walther von der Vogelweide Fouqué   Friedrich II. von Preußen
                        Weber   Freiligrath
                                       Kant   Ernst   Frey
Fechner   Fichte   Weiße Rose   von Fallersleben   Richthofen   Frommel
                Engels   Fielding   Hölderlin
        Fehrs   Faber   Flaubert   Eichendorff   Tacitus   Dumas
                    Maximilian I. von Habsburg   Eliasberg   Ebner Eschenbach
    Feuerbach   Fock   Zweig
                        Ewald   Eliot   Vergil
            Goethe   London
    Mendelssohn   Balzac   Shakespeare   Elisabeth von Österreich   Dostojewski   Ganghofer
                        Lichtenberg   Rathenau   Doyle   Gjellerup
        Trackl   Stevenson   Hambruch
    Mommsen   Tolstoi   Lenz   Droste-Hülshoff
            Thoma   Hanrieder
    Dach   Verne   von Arnim   Hägele   Hauff   Humboldt
            Reuter   Rousseau   Hagen   Hauptmann   Gautier
        Karrillon   Garschin   Baudelaire
            Damaschke   Defoe   Hebbel
                            Descartes   Hegel   Kussmaul   Herder
Wolfram von Eschenbach   Dickens   Schopenhauer   Rilke   George
        Bronner   Darwin   Melville   Grimm Jerome   Bebel
            Campe   Horváth   Aristoteles   Proust
    Bismarck   Vigny   Barlach   Voltaire   Federer   Herodot
                        Gengenbach   Heine
        Storm   Casanova   Tersteegen   Gilm   Grillparzer   Georgy
                    Chamberlain   Lessing   Langbein   Gryphius
    Brentano   Lafontaine
        Strachwitz   Claudius   Schiller   Kralik   Iffland   Sokrates
                Katharina II. von Rußland   Bellamy   Schilling
                                    Gerstäcker   Raabe   Gibbon   Tschechow
    Löns   Hesse   Hoffmann   Gogol   Wilde   Gleim   Vulpius
        Luther   Heym   Hofmannsthal   Morgenstern
                Roth   Klee   Hölty   Goedicke
        Luxemburg   Heyse   Klopstock   Puschkin   Homer   Kleist
            Machiavelli   La Roche   Horaz   Mörike   Musil
    Navarra   Aurel   Musset   Kierkegaard   Kraft   Kraus
                Nestroy Marie de France   Lamprecht   Kind   Kirchhoff   Hugo   Moltke
            Nietzsche   Nansen   Laotse   Ipsen   Liebknecht
                        Marx   Ringelnatz
            von Ossietzky   Lassalle   Gorki   Klett   Leibniz
                        May   Lawrence   Irving
                                vom Stein
        Petalozzi
                Platon   Knigge
            Sachs   Pückler   Michelangelo   Kock   Kafka
                    Poe   Liebermann
                    de Sade   Praetorius   Korolenko
                                Mistral   Zetkin

The publishing house tredition has created the series **TREDITION CLASSICS**. It contains classical literature works from over two thousand years. Most of these titles have been out of print and off the bookstore shelves for decades.

The book series is intended to preserve the cultural legacy and to promote the timeless works of classical literature. As a reader of a **TREDITION CLASSICS** book, the reader supports the mission to save many of the amazing works of world literature from oblivion.

The symbol of **TREDITION CLASSICS** is Johannes Gutenberg (1400 – 1468), the inventor of movable type printing.

With the series, tredition intends to make thousands of international literature classics available in printed format again – worldwide.

All books are available at book retailers worldwide in paperback and in hardcover. For more information please visit: www.tredition.com

tredition was established in 2006 by Sandra Latusseck and Soenke Schulz. Based in Hamburg, Germany, tredition offers publishing solutions to authors and publishing houses, combined with worldwide distribution of printed and digital book content. tredition is uniquely positioned to enable authors and publishing houses to create books on their own terms and without conventional manufacturing risks.

For more information please visit: www.tredition.com

Wild Ducks How to Rear and Shoot Them

W. Coape Oates

Imprint

This book is part of the TREDITION CLASSICS series.

Author: W. Coape Oates
Cover design: toepferschumann, Berlin (Germany)

Publisher: tredition GmbH, Hamburg (Germany)
ISBN: 978-3-8491-6534-5

www.tredition.com
www.tredition.de

Copyright:
The content of this book is sourced from the public domain.

The intention of the TREDITION CLASSICS series is to make world literature in the public domain available in printed format. Literary enthusiasts and organizations worldwide have scanned and digitally edited the original texts. tredition has subsequently formatted and redesigned the content into a modern reading layout. Therefore, we cannot guarantee the exact reproduction of the original format of a particular historic edition. Please also note that no modifications have been made to the spelling, therefore it may differ from the orthography used today.

WILD DUCKS

"*The Fleet at Flight time.*"
W.L. Colls. Ph. Sc.

TO

MY WIFE

PREFACE

The main object of this book is to assist those who are anxious to rear wild ducks on economical lines. The Author is not without hope that the pages which it contains may even be of some use to old hands at the game.

CONTENTS

CHAP.
I. SELECTION OF STOCK AND THEIR HOME
II. LAYING AND SITTING
III. HATCHING AND REARING
IV. SHOOTING

LIST OF ILLUSTRATIONS

PHOTOGRAVURE PLATES
From Drawings by G.E. Lodge

THE FLEET AT FLIGHT TIME

ON GUARD

A TIDY MOTHER

QUITE TALL ENOUGH

FROM PHOTOGRAPHS

COMING IN TO FEED	"
THE CAGE	"
THE REARING PADDOCK	"
A SMALL RUN	"
WARE WIRE!	"
WIRED IN ON THE WATER	"
AN INEFFECTIVE CRIPPLE STOPPER	"
BEFORE THE EVENING MEAL	"
A RIGHT AND LEFT	"
AT THE END OF THE DAY	"
COMING ON A SIDE WIND	"
LADIES IN WAITING	"

   SELECTION OF STOCK AND THEIR HOME
   [Pg 13]

WILD DUCKS

CHAPTER I

SELECTION OF STOCK AND THEIR HOME

The first point to be decided by the would-be owner of wild-fowl is the locality where he intends to turn down his stock.

Wild-fowl can undoubtedly be reared far from any large piece of water, but I am strongly of opinion that birds do better on a good-sized stretch of water with a stream running into it and out of it. Given these advantages, the running water must be constantly bringing a fresh supply of food, especially after a fall of rain sufficiently heavy to cause a rise of water; further, if the stream which runs out of our lake empties itself into a large river, the latter will, when it floods or rises rapidly, cause [Pg 14] our stream to back up and bring in a further supply of food from the main river.

Some morning the ducks are absent from their accustomed haunts, and if we walk up to the spot where the stream enters the lake, ten to one we shall find our birds there thoroughly enjoying some duck-weed or other food swept down by a rise in the water.

This supply of fresh food is a gratifying source of economy to the grain bill at the end of the year, and it is most fascinating to watch the birds "standing on their heads" in their endeavours to reach this change of diet.

Another great advantage, too, is that a far higher percentage of fertile eggs will be obtained if the ducks have a large piece of water at their disposal.

Given these advantages, it is, however, most necessary for the birds to have some shelter near the lake, both as a protection against the weather and to serve as suitable nesting places.

Nothing, for instance, could be better than a stackyard or paddock in the vicinity [Pg 15] of the water, and if the paddock is bounded by a flood bank or tall hedge, giving shelter from the prevailing wind, so much the better.

Ducks love to nest in stacks, and I have known a pinioned bird work her way up the side of a stack and make her nest fifteen feet from the ground. In stacks birds can burrow so deep that no weather, however inclement, can damage the eggs.

Outhouses too are very favourite places for ducks to lay in; also old stick heaps and the bottom of thick hedges. My main point is this, that if you take the trouble to regularly feed your wild ducks morning and evening and keep them quiet, you will soon find that you can get them *to lay where you want them to lay*, and the places you select will naturally be those where they are secure, or nearly so, from their natural enemies, such as rats, cats, weasels, moles, and other vermin.

This is the first secret of success.

I have seen wild ducks so tame that within a fortnight from the time they first joined my own birds they were eating maize close to my feet.

[Pg 16]

Having obtained my piece of water and decided on the spot where I mean to feed my birds, the next step is to get the breeding stock.

I consider that the best time to purchase the stock is December, as this gives ample time for the birds to pair and get used to their surroundings before the breeding season commences; one is almost sure to get some cold weather in January, and the cold will make the birds more dependent on the food given to them, and therefore more easily managed.

Next as to the stock and where to get it.

I advise you to obtain your birds from different places, two or three birds from each place, taking care to get fairly young birds, and not older than, say, two years. By this means you will get a certain amount of change of blood, particularly during the second season, when the different broods, which have been well mixed at hatching time, pick their mates and breed.

COMING IN TO FEED

I believe that this method is more satisfactory than buying eggs in the first instance, as in the latter case you cannot tell for certain how long the eggs you purchase have been laid, nor what the birds are like which laid them. We next come to the question of the proportion of drakes to ducks. On a small piece of water, one drake to every three ducks will do very well, but if you have at your disposal a large lake, I am strongly in favour of plenty of drakes, say fifteen drakes to every twenty ducks. Most of the birds will pair, though occasionally one finds as many as three drakes paying court to one duck, and one drake taking away two or even three ducks.

It will generally be found, however, that if any of your ducks are without mates, wild birds will soon come and pair with them, and this is, of course, just what you want. I have adopted this principle for some time, and practically all the eggs collected are fertile. It will be found that at times — particularly whilst the ducks are sitting — the drakes are a great nuisance, but at this period one can always catch them and shut them up.

The next point to be considered is as to what food is best for breeding birds, and I say unhesitatingly maize. There is practically no waste, and you have not the mortification of seeing crowds of sparrows swoop down on your ducks' food as you turn away.

Better still, ducks lay capitally on maize, and you may calculate on obtaining an average of twenty-three to twenty-four eggs apiece from your ducks if fed carefully.

You will find that strange ducks when they first join your own will not eat maize, though they soon take to it when they see your own birds feeding.

It is easy to tell the advent of a stranger by this peculiarity, and by his generally alert and suspicious manner.

I am a strong believer in the infusion of fresh blood each year, and this is easily done by catching a few stranger drakes and pinioning them. These birds, if kept up until their wound is healed, and then enlarged in good time, will pair with your own birds and often become very tame. I did not find that pinioning strange ducks answered so well, as they were very prone to stray and lay their eggs at a distance, and their young were always shy and difficult to tame; moreover, [Pg 19] the ducks never bred the first year after pinioning, whereas the drakes did. It is quite a simple matter to catch these wild birds; you have only to construct an ordinary wire-covered cage, somewhere near the water, and with the face nearest the water closed by a door; you then accustom your own birds to feed inside this cage, and you will soon find that in winter they will come for food as soon as it is light, or rather just as day is breaking, always provided that you feed them at that time.

You have been careful to leave the door of the cage open over night, and have put some maize inside the cage. A strong cord attached to the door is passed across the doorway and round a wooden "runner" on the opposite post, and then to the back of the cage, where your man lies concealed. Often during severe weather, which is always the best for this kind of work, your own birds will be followed by one or two strangers, who in the half light come inside the cage before realising their mistake. Once you get them inside the cage with their heads away from the entrance, pull the string and [Pg 20] shut the door. Care should be taken that the string is fairly high up, so as not to catch the duck's eye. Having got your birds safely inside, catch them quietly and quickly, and having pinioned them, take them, if possible, to a cage with some part of it projecting out into the water. You, of course, feed them regularly, and are careful to give them some artificial cover to skulk in, as for some time the pain of the wound and the fright they have had makes them terribly shy. This cage, once constructed, is most useful for such work, and can be built at trifling cost, and the size I would recommend is about fifteen yards long by five yards wide, with a height of five or six feet. Your own birds soon get used to their part of the business, and, if you are quiet and quick, soon get over their nervousness.

The advantage of confining your captives for a short time is obvious. They get used to their surroundings and recognise the lake as their new home, and soon take to their diet of maize, so that when you liberate them they rarely give much trouble, and readily mate with your own birds.

THE CAGE

[Pg 21]

One very important point which I have omitted to mention is the necessity to kill down all rats, hedge-hogs, moles, and weasels in the vicinity of your breeding places. Rats are the ducks' worst enemies, and I have known one old doe rat which had no less than sixteen wild ducks' eggs in her larder when she was dug out and killed. All these eggs had a small hole in them, and were of course spoilt. We proved conclusively that she had no partner in her crimes, as we never lost another egg after her death. Rats are a perfect curse to young ducks, and they will carry them off even when they are half-grown, occasionally killing two or three ducklings in a single night without even taking the trouble to remove them. On another occasion I remember a rat killing a duck whilst sitting on her nest; the unfortunate bird had allowed herself to be killed apparently without moving.

Moles do a good deal of damage by burrowing under the nests, thus forming a cavity into which the eggs fall; they are then carried off by the mole. More than this, many a duck is either put off laying [Pg 22] or induced to desert her nest when sitting owing to the restless movements of this little pest.

A last word as regards the numbers you should retain as a breeding stock. This largely depends on the size of the piece of water you own and the amount of food it can supply to your birds. If your stock is too large, your birds will do a lot of harm to the meadows adjoining the water, and you must bear in mind that the possession of the goodwill of the farmers round is the second secret of success. Ensure this, and you don't get eggs stolen, and, better still, you are informed of the whereabouts of any truant ducks that may be nesting away from home.

A present of a couple of fat wild ducks will cover a multitude of their sins.

[Pg 23]

LAYING AND SITTING

[Pg 24]

"*On Guard.*"
W.L. Colls. Ph. Sc.

[Pg 25]

CHAPTER II

LAYING AND SITTING

We now come to the time when the ducks, having paired, show an inclination to look for suitable nesting places. The drake takes the lead in this, and you may be sure that when you see birds peering about in hedge bottoms, stick heaps, &c., that eggs will soon be laid.

At this time, too, they use a different note, and to quote a very apt term used by a friend of mine, they "begin to talk." About the beginning of February it is advisable to hint to the ducks where you want them to lay. If you have any large trees in your paddock, place a number of sticks up against the trees in the form of a circle, leaving one or two clear spaces inside the heap. Then make some circular holes, one in each of the spaces, and about five or six inches deep, and shelving gradually from rim to centre. It [Pg 26] is best to scatter some sand in these holes, so that the birds can more easily work the nests to the dimensions that suit them. Don't make the nests too small or too shallow, as they may have to contain fourteen or fifteen eggs. It is advisable to put some short dry grass or old hay near the nest, and a very little in it, so that the duck can manipulate it at her pleasure.

The principal thing to remember is, that the nest must be sheltered as much as possible from draughts, and be made well in the middle of the cover, as ducks like darkness when they are sitting. Broom is about the best cover you can use for sheltering a nest, and is most adaptable. Practical experience, and one's early failures, teach one more than anything else how a nest should be made, and yet often when you are satisfied that you have selected a most suitable spot for nesting purposes, you will find a duck occasionally preferring a miserably draughty position for her nest within a yard of the snug retreat you have devised for her. The only thing then to be done is to leave her alone until she has settled down to lay [Pg 27] steadily, when you can gradually introduce pieces of broom, &c., so as to shelter her nest as much as possible from wind and rain, taking care to leave the entrance to the nest clear. Young ducks as a rule are the most shy, and you will generally find the older birds only too glad to avail themselves of the well-sheltered nests that you have provided for them.

Nothing can be better for ducks to nest in than the corners of an outhouse or old stable, always provided that you have killed off the rats.

In such places wind and rain can do no harm, and practically every egg hatches out.

The roots of hollow willow trees are favourite nesting places, but a bit dangerous if too near the water's edge. Many birds delight in straw stacks, and if disturbed will simply go up higher, so as to be out of the way of cattle or human beings.

I believe that if you can get your birds to nest in outhouses or stacks, you will get a much better hatch out than elsewhere. Last year one of my ducks took off all her sixteen eggs safely from the corner of a [Pg 28] stable, and a bird sitting close to her hatched eleven, without a single bad egg; and we had almost as good results from birds nesting in stacks.

One bird, after being disturbed from her nest in the side of a stack, built at the top, and quite twenty feet from the ground. One fine morning we found her with fourteen young ducklings, and she appeared much annoyed at the assistance which we gave to the family to descend.

If the weather is dry and your nests are well situated, your birds nesting outside may do as well as those described above; but given a week of cold wind and penetrating wet, down goes your average at once.

Last season was a particularly favourable one, and from the first five nests (all sat upon by ducks) no less than sixty-five ducklings hatched out—a highest possible. Naturally this extraordinary percentage was not maintained. We will now suppose that the ducks have begun to lay, an event which may take place any time from the middle of February to the middle of March, after which date they ought to be laying steadily. [Pg 29] As they will lay many more eggs than they can successfully hatch, pick up some eggs at intervals from the nests, taking care always to leave two or three in each nest. These eggs should be placed on a large tray or shallow box, lined with hay, sawdust, or other suitable material. It is not advisable to place them touching each other, and care should be taken to turn them daily; if this is done the eggs will keep well for three weeks, by which time you have collected a sufficient number to put under hens, however small your stock may be.

Eggs left in the nest will, of course, not require turning, as the duck does this herself.

When you have collected a number of eggs, place them under hens, having first satisfied yourself that the hens are good sitters. Eight to ten sittings of twelve eggs each is a good number to put down as a start, as from this number you ought to get about a hundred ducklings, and these, when old enough, can be divided into two runs of about fifty each. I have found by experience that it is unwise to put a larger number than this together until the birds are about six or [Pg 30] seven weeks old. Naturally, the number of eggs you can put down will depend on the size of your stock and the number of sitting hens at your disposal.

A certain amount of care is necessary in preparing the nest for the hens, as ducks' eggs are very fragile, and much more easily broken than hens' eggs.

The following is the method which I recommend. Get any square box of sufficient depth, and having cut some pieces of sod, build up the corners of the box with them: then cut a square sod to fit the size of the box, and having removed some of the earth underneath the centre of the sod, place it grass upwards in the box. By this means you will obtain the proper shape for the nest, viz., a gradual slope down from the sides to the centre; this will prevent your hens accidentally kicking eggs from under them, as owing to the shape of the nest any eggs which are displaced must roll towards the centre or lowest part of the nest; there is consequently little danger of any of the eggs getting cold. After this, line the nest with dry moss. The sod underneath has [Pg 31] the advantage of producing greater heat, and gives a more satisfactory hatch out than nests made of other material, and being firm does not lose its shape.

Don't forget to give your sitting hen some ventilation, but be careful that *no draught can reach the eggs.*

The sitting hens will, of course, be taken off to feed regularly every day, and you will find them give you less trouble if you take care to tether them on the same leg each day.

And now to return to the laying ducks.

As time goes on you must leave more eggs in the nest, as the birds will soon want to sit. A duck shows signs of this by lining her nest with down from her breast, and in a short time you will find the whole nest, sides and bottom, lined with a thick covering of down; while the eggs are covered by what I can best describe as a thick movable quilt, which protects them from the cold, and the prying eyes of carrion crows and other poachers.

At this time you will observe the old duck staying longer and longer on her nest each day as she lays the last egg or two, and [Pg 32] you may be sure that she has fairly begun to sit if you find her still on her nest about 6 or 7 P.M. A day or two before she begins to sit, her nest should be made up to its proper complement of eggs, and it is always wise to keep a few eggs in hand for such contingencies. The number of eggs a duck can sit on depends largely on the size of the duck and also the depth and breadth of the nest; given favourable conditions a duck can manage sixteen or seventeen eggs, and I knew of one nest, consisting of sixteen eggs, all of which hatched off. There is, however, this risk, that should bad weather come it is practically impossible for a duck to successfully brood so large a number as sixteen ducklings, even when her coop is turned away from the wind and rain; and it is here that large brooding hens such as the Bufforpington score their strongest point as mothers to young ducks.

Of one thing you may be sure, a duck will not retain any more eggs in her nest than she can conveniently cover. I know of one case where a duck belonging to me was sitting on fifteen eggs. All appeared to be going well, until one morning a friend of mine, on whose veracity I can absolutely rely, saw the duck fly from her nest, close to where he was standing, *with an egg in her bill.*

"*A tidy Mother.*"
W.L. Colls. Ph. Sc.

[Pg 33]

She flew to the water, about 150 yards away, apparently without breaking the egg; but unfortunately my friend could not get up in time to see what she did with it. She hatched out the rest of her eggs satisfactorily.

I presume that either the egg in question was cracked and she removed it for the sake of cleanliness, or because she felt herself unable to sit on so many eggs.

On many occasions I have noticed an egg left bare on the top of the downy covering which ducks are so careful to leave over their eggs when they go off to feed, and these eggs, if taken away and placed under a hen, have invariably hatched. To the best of my recollection I have never known eggs disappear from a nest containing eggs up to thirteen in number; but over that I could quote many instances of one or two eggs going.

This has led me to believe that the bird above alluded to had removed an egg from [Pg 34] her nest, as she felt herself unable to sit on so many. A good number of eggs to leave under a duck is thirteen, and under a hen twelve.

I have satisfied myself that hens, however small and light, break many more eggs than ducks, and for this reason I do not care to give a hen too many—one broken egg frequently leads to more.

It is advisable when once the ducks have begun to sit, to catch their mates, if possible, and shut them up in some convenient place during incubation, as otherwise they bully the sitting ducks when they come off to feed, and you may have the annoyance of seeing a duck desert her nest just at hatching time, as nature has warned her that she must shortly lay again. I had one instance of this kind, when a duck which had been sitting very steadily left her nest when the eggs were actually "spretched" (cracked previous to hatching), and as later in the day she showed no signs of returning we had to put them under a hen. The duck in question never returned to her nest, but soon made another. She had not been disturbed in any way.

[Pg 35]

Should a duck forsake its nest, place the eggs under a good hen as quickly as possible, even if they are stone cold.

I had one case last year, which I thought hopeless. The eggs had been sat on for about a fortnight. They were stone cold, and we knew the duck had been off her nest for at least twelve hours, probably much longer. Eventually twelve out of the thirteen hatched. If you are unable to catch the drakes, the best plan is to put food and water near the nest of the sitting birds, the pan containing the water being large enough to allow her to wash herself thoroughly, as it is the daily tub which generates heat, and assists most materially the successful hatching of the young birds.

I will now deal with the vexed question as to the best kind of hens to be employed. Personally I have strong leaning towards "Bufforpingtons"; they are, of course, heavy, and do break a few eggs—ducks' eggs being particularly brittle—but, on the other hand, they are very staunch sitters, quiet and easy to handle, and not likely to get excited when other hens are hatching in close [Pg 36] proximity to them. I have tried lighter hens of several breeds, and I find that they break as many eggs, and trample on as many young ducklings as the Buffs, whereas taking them all round, they are not so easy to handle, do not sit so steadily, and have nothing like the wonderful brooding capacity of the Buffs.

Many people put all their wild ducks' eggs under hens, and do not allow the ducks themselves to sit. I think this is a mistake, as nature gives to ducks far greater powers to hatch their own eggs than she gives to hens. The daily bath, already alluded to, and the mass of warm soft feathers, greatly assist in generating heat, and in preventing the eggs from getting chilled.

The old duck treads more lightly when going on to her nest, and certainly breaks far fewer eggs than the hen does. On the other hand, ducks are not such good "brooders" as hens, and are far more likely to get dirty when kept under coops, however often you may change the ground, owing to the fact that they do not get to the water for the daily bath which is essential to [Pg 37] them; and if you leave a bath for them in the coop, the young ducklings will be sure to get to it and probably contract cramp.

Another strong point in favour of hens is the fact that when you have a large number of cletches of ducks in the wired run the hens do not kill them when they make a mistake and go to the wrong coop, whereas ducks frequently do. If, therefore, a considerable number of broody hens are available, the best plan is to let the ducks sit on the eggs until they are "spretched" (cracked), and then transfer them to hens which have been sitting for some time. This, however, is a cruel business at best.

The plan I always adopt is to note down carefully the day on which a duck should hatch, and having satisfied myself that the young ones are dry after hatching and ready to move, I catch the old duck on the nest, and remove her and her whole family to a coop and run.

Care of course should be taken to see the bars in front of the coop are not sufficiently far apart to allow the duck to escape.

[Pg 38]

Ducks' eggs take from twenty-four to twenty-nine days to hatch as a rule, though occasionally a lot of eggs that have been put down soon after being laid will hatch in twenty-three days, if set under a good hen. I should put twenty-six days as the usual period of incubation.

If the ducks are well and regularly fed, they should lay an average of twenty-three eggs apiece during the nesting season. We generally feed ours on maize, as it is less wasteful than smaller grain, and the birds lay well on it. One can, I think, count on 80 per cent. of the eggs hatching, and of birds actually hatched you ought, in a fair season, to rear 85 per cent. Having taken my reader as far as the hatching out of the young birds, I propose in my next chapter, which I consider the most important in the book, to deal with the question of their food, up to the time they are fit to shoot.

[Pg 39]

HATCHING AND REARING

[Pg 40]

THE REARING PADDOCK

CHAPTER III

HATCHING AND REARING

The time is now approaching when the ducklings may be expected to hatch out. Care has been taken to plentifully sprinkle the eggs with tepid water, two or three times a week, whilst the hens are off for their daily feed, and everything is ready for the young birds.

The first sign of approaching hatching is a curious opaqueness which affects the eggs. This is speedily followed by chipping, and by placing the egg close to the ear the young birds can easily be heard endeavouring to obtain their liberty. If all the eggs chip and hatch together it is a sure sign of healthy birds; but should they be hatching out patchily, remove the earliest birds at intervals from the different hens and put them in a basket lined with flannel, in a warm corner of a room, but not too near the fire.

When the birds are quite dry and lively, remove them eleven or twelve together and place them in a coop, with a small wire run attached. Always place the coops facing the sun, if possible, and with their backs to the wind.

Wind and rain must be kept out and the sun admitted. The latter will bring on young birds quicker than anything. During very bad weather coops may have to be shifted two or three times a day if the wind keeps changing.

A matter of considerable importance is the nature of the soil on which the ducks are to be reared. Let it be light and well drained, and the ground undulating, so that it may be always possible to shelter the young birds from a harsh wind. A high bank, such as that alluded to in Chapter I., is often of the greatest assistance in sheltering them, particularly when they get a little older.

The ducklings must be put on some grass, as otherwise the hens will scratch for food, and generally damage one or two of their brood. The young hopefuls require scarcely anything to eat or drink for the first twenty-four hours, and do little else [Pg 43] but brood underneath the hen, though little patches of brown and yellow with a bright eye here and there form a fascinating picture for any passer-by.

The first food given should be a little fine wild duck meal, scalded in the usual way, and put on a *shallow* plate outside the coop, and inside the small wire run attached to the coop. To start with, a little food may be scattered over the grass inside the coops to attract the little birds to their meal; they should, however, be taught to feed as soon as possible outside for the sake of cleanliness.

It is most important at this early stage that the food be not of too sloppy a nature, otherwise the birds soon get in a terrible state, and absolutely coated with their food. This always leads to their heads, eyes, and often their backs becoming sticky, and in the end spells a big death roll. Very little water, and that pond water, should be given during the early stages; the colder the weather the less they ought to have to drink, and it is often a good plan to take the chill off what little is given them. Don't forget to give the hens food and water twice a day.

[Pg 44]

A busy time is now before the keeper, or whoever is responsible for the feeding.

The earliest meal should be given at about 4 A.M. or as soon as it is light, and then regularly throughout the day, every four hours.

Be careful to feed more frequently during inclement weather, and move the coops prior to feeding.

The ducklings are now fairly started on their journey, and before following them I propose to describe the method to be adopted in the case of ducks sitting on their own eggs. Visit the sitting birds pretty regularly, so that they lose their shyness on seeing you, and when the birds have been sitting twenty-five days, go in the early morning and late in the evening to satisfy yourself that the eggs have not hatched.

An experienced eye can generally tell, by the unusually elevated appearance of the duck on her nest, when she has hatched, and sometimes by creeping quietly forward the little birds may be heard chirping, though they instantly cease on receiving a warning from their mother.

[Pg 45]

Should you have doubt as to the hatching having taken place, a blunted stick put under her breast will generally reveal the state of affairs, and if she knows you the old bird will not mind this.

Directly it is ascertained that the brood is dry enough, the old bird should be caught and the little ones put in a warm basket, and the lot transferred to a coop and run, after which they are treated in exactly the same manner as those under a hen.

I do not, however, think it wise to give the duck as many young ones to look after as are given to the hen; very often, however, there is no option in the matter.

Ducks are very cunning at hatching time, and unless the keeper is quick and observant, will frequently give him the slip, and get off with their brood to the water, where more than half of them will probably fall a prey to rats and pike.

I remember on one occasion being unable until late afternoon to go and look at a nest which was due to hatch in the morning. I found fourteen empty shells and the family gone. It was a very cold day, and after a [Pg 46] prolonged search the party were found snugly ensconced under a willow tree. They were speedily captured and brought home.

The young ducks are now three or four days old, and have got into the way of running out of the coop and into the run for their food and water. They have overcome their early shyness, and on the appearance of the keeper speedily show themselves. A little fine crissel and flint grit can now with advantage be added to the meal, and some sand, which acts as a digestive, placed in the water and on the grass. Never give them more than they can eat. Nothing is worse than stale food left about; it leads to diarrhœa, &c., and gives the youngsters a distaste for their food. The food can be placed in long shallow troughs or on the grass in one long line. I prefer the former plan, as less is left about to become stale and sour. Care should be taken to see that the troughs are thoroughly washed after each meal.

When about ten days old the ducklings require more room to roam about in, and unless you give it them they will begin to go back.

A SMALL RUN

[Pg 47]

Place five or six coops, hens inside, in a line, and about a foot apart, and wire in a piece of ground about ten yards square round the coops; it is better to give them too much room than too little. It will generally be necessary to move on to fresh ground every four or five days during this stage, but much depends of course on the state of the weather. It is a good plan to leave the small wire runs inside the larger runs, as they give a certain amount of shelter in bad weather. It is delightful to see the little chaps appreciating their new liberty and dashing about in all directions in chase of flies, &c. Nothing seems to hurt them at this time, and I once remember seeing three of my young ducks devour a bee apiece after first crippling it. I have noticed a bird swallow a bee alive, and have also seen one stung, but no ill effects resulted.

It is a good plan now to give the birds a little boiled rabbit, chopped up fine; it makes a change from the crissel, and ducklings must have some animal food as a substitute for the slugs, worms, and many etceteras that they pick up in their natural [Pg 48] state. The chopped rabbit should be mixed with the meal.

One word of advice before going farther. Previous to placing a number of coops containing the old ducks close together, ascertain carefully whether there are any vicious ones amongst them—some are very savage, and will immediately peck to death any unwary little one which enters a coop not its proper home. It is best in these cases to isolate the old bird and her brood altogether, if you have plenty of room, or, failing that, to place her by herself in one corner of the run.

If bad weather comes on, a pinch of "Cardiac" (a kind of tonic sold by Messrs. Spratt) may be added to the food, but I only advocate its use occasionally. The chief point I am anxious to impress on my readers is, don't let your birds get cold and wet; if you do, ground is lost which can never be recovered. A capital plan is to cover some portion of the run with sacking or a waterproof sheet to form a shelter against excessive heat or a sudden hail-storm. The most delicate time, in my opinion, is just when they are getting their shoulder feathers, and if you get them safely through this period the worst is over.

WARE WIRE!

[Pg 49]

When they are about a fortnight old begin to give them some wheat in their drinking water; that known to farmers as "seconds" is best. I am a strong advocate of steeping the wheat before feeding, as I think it renders it more digestible, though this is not so necessary if one uses "seconds." The ducks having got to eat wheat nicely, introduce a little barley, and by the time they are seven weeks old you can afford to do without meal entirely, and it will be time to take the birds down to the water which is to be their home. The greatest obstacle to success in rearing during the early stages of a young wild duck's life is the extraordinary knack they have of getting their heads and backs dirty. This is a most serious matter, and causes great mortality unless attended to. It is generally caused by the food adhering to their heads and cheeks; being of a sticky nature, it will often, if neglected, cause inflammation to the eyes and eventually blindness. If once their [Pg 50] heads get dirty, their backs soon follow suit, as the act of "preening" soon transfers the dirt from the head to the back.

This curse to young ducks is most prevalent in wet weather, and it is therefore most necessary to constantly change the ground so as to keep the birds as clean as possible; if once the old bird gets dirty, it is good-bye to the general good health of her brood.

The only remedy, if matters become serious, is to get some tepid water and soap, and carefully wash the affected places with a soft sponge, taking care to free the down or feathers which have adhered to the skin. A hot sunny day is the best for the purpose, as the young birds then have every chance of getting dry. If the old bird is dirty, try to allow her a good wash in a tub or small tank; she must, however, be watched, otherwise she may leave her young ones in the lurch. If your ducks are pinioned it is easy to manage this bath, and to prevent the birds straying afterwards from their young. When the ducklings are seven weeks old choose a nice warm day, and take them down to the water: I say a warm day, as owing to their delight at getting to their natural element, they are very liable to overdo their bathing at first, and, should the day be cold, the casualty list will be a big one next morning.

WIRED IN ON THE WATER

[Pg 51]

At this time it is best to wire off a piece of land and water, making the whole into one large run, and taking care that there is some shelter on land for the young birds. It is a good plan to bring down the old birds, coops and all, to their temporary home, keeping the mothers shut up in the coops for the present. Their presence gives confidence to the ducklings, and their sharp warning "quacks" tell them when danger is about, and also emphasise the fact that there do exist such things as gulls, carrion crows, cats, dogs, &c., and that in future the young hopefuls must look out for themselves. Willow trees planted at the water's edge and kept about five to six feet high form admirable protection from bad weather and winged vermin, and also give welcome shelter from the heat of the sun, whilst they undoubtedly add to the amount of insect life in the run.

[Pg 52]

If you wish to study economy in feeding, an excellent plan is to mix barley meal with your duck meal; commence in the proportion of four parts duck meal to one of barley meal, and increase the proportion of the latter until the mixture is half and half. Too much barley meal is, I feel sure, a bad thing, and causes indigestion, and if expense is no object it is best to stick to the wild duck meal until the ducks are weaned to corn; if, however, you do decide to feed on barley meal, it is a good plan to mix a little bran with it, in the proportion of one part bran to four of barley meal.

You should get the ducks on to corn as soon as possible, and teach them to eat it in shallow water; they don't eat it so fast if this plan is adopted, are less liable to get indigestion, and in searching for the food are constantly drinking water at the same time as the food, as well as a certain amount of grit, sand, &c. Ducks must have water with their food, and the sooner they are trained to take corn and water together, the better will their meals be digested; moreover, if fed in this way birds certainly [Pg 53] require less, and there is consequently a gratifying reduction in the grain bill at the end of the year.

To return to the run at the water's edge, let it be assumed that the birds have been ten days to a fortnight in their new home, have become thoroughly accustomed to it, and naturally look upon it as the place where food is to be obtained at stated times.

It is now time to enlarge them altogether, but before doing so liberate a few of the least vicious of the old ducks. These birds very soon take charge of a certain number of young ones, and directly the wire is pulled up will teach them where to look for food. It is a very pretty sight to see an old bird swimming at the head of twenty or thirty young ducklings, who form a compact mass behind her, and always accompany her in foraging expeditions. She it is who warns them that it is nearly feeding time; it is her eye which has detected a well-known figure hovering overhead, and her voice which warns them to make for the nearest shelter.

By this time I am sure that my readers will be getting impatient because I have said [Pg 54] so little as to the cost of food. A golden rule is to give your ducklings all they will eat during the first seven or eight weeks, and after that make them hunt for their natural food, giving them just sufficient to keep them fairly fat and prevent them from straying. It is quite possible to get them fat enough for the larder by increasing the supply of maize during the last fortnight or so before your shoot takes place.

I am of opinion that, provided a man feeds and looks after his ducks himself, is in possession of a supply of coops and runs, and is fortunate enough to have a suitable piece of water of his own, as well as a bit of ground to rear them on, that he can make his accounts balance at the end of the year. In other words, he will be able to give his friends some very enjoyable shooting, and supply himself with a hobby of which he will never be tired, at no expense to himself. In support of my statement I propose to give a few figures. The breeding stock has of course to be purchased, and for the sake of simplicity let us put it at twenty ducks and fifteen drakes, making an initial cost of [Pg 55] about £7. In an experience of some years, however, I have found that my stock at the end of the season numbers practically the same as at the commencement, and I found it always possible to fill up any casualties by catching and pinioning wild birds which join my own. On these grounds I consider that my stock at the end of the season is of the same value as at the beginning, and that one side of my account balances the other.

The stock fed on maize will cost about 12s. 6d. a month, and, supposing that the first birds are hatched out about the middle of April, and practically all, except a very few retained for breeding purposes and some immature birds, are killed the first week in September, this calculation brings the price of feeding the breeding stock for seven and a half months to £4, 13s. 9d.

Now for the food of the young birds. I assume that from the above-mentioned stock about 250 ducklings will be reared, and, taking an average of several years, their food from the date of hatching (mid April) to early September works out roughly at £16. This includes wild duck meal, wheat, barley, [Pg 56] and barley meal, a little maize, and the many etceteras, such as crissel, grit, and cardiac.

To this should be added a little extra for the feeding of the immature birds, which are not quite ready for killing. Put this at 15s.

In addition there is still the expense of sitting hens: if twenty hens are purchased at 3s. each and afterwards sold at 2s., this item will work out as an expenditure of £1. They have of course to be fed, but their food—maize is the best—has been taken from the food purchased for the ducks, so that no further amount has to be charged under this heading.

The debit side of the account will now work out as follows:—

Food for ducklings	£16	0	0
Food for old birds	4	13	9
Extra food alluded to above	0	15	0
Expenses for sitting hens	1	0	0
	£22	8	9

As against these figures there are 250 young ducks for sale: deduct from this number fifteen for casualties of various kinds, such as dead birds unpicked at the [Pg 57] shoot, odd birds that may stray and be killed, &c., and this gives 235. If the birds are properly fed a game-dealer will be glad to give 2s. each for them, especially if the shoot is timed to fit some popular function, such as Doncaster Races; so that the credit side of the account shows a sum of £23, 10s. for the sale of 235 birds, giving a small surplus of rather over £1, which can be used to meet incidental expenses, such as purchase of wire, &c. Each young bird will cost about 1s. 3½d. to rear, and will sell for 2s., leaving 8½d. a bird profit with which to meet the other expenses. Many of my readers may think the margin of fifteen birds set apart as casualties far too small, but I can assure them that, so far with me, it has never reached that number, and need not do so provided the birds are kept at home by proper feeding, and the right people propitiated.

Naturally one does not sell all one's ducks, or anything like it. Some are given to the friends who come to the shoot, and many are given to the farmers round, but in considering accounts, I think I am justified in including the value of birds given away as [Pg 58] one of the assets. In any case I have made an honest attempt to help those who wish to look before they leap. Ducks are very fond of maize; it certainly brings them on quicker than anything else, and I have had young drakes of the year in full plumage on August 1, when maize has been the only corn used. It is, however, too fattening, I think, and a bit apt to make the birds lazy. I do not believe that birds fed solely on maize fly so well or are as good for the table as those whose diet is composed of a mixture of wheat, barley, and maize. The birds must be encouraged to seek their natural food, as only by this means will the wild duck's flavour be retained.

The birds must be fed at regular hours, as this is the only guarantee that they will be at home when wanted.

I hope that in this chapter I have succeeded in showing how wild ducks can be fed in the best and also most economical manner, and I shall endeavour in the concluding one to give my readers some hints as to how the birds can be made to show reasonably good sport.

[Pg 59]

SHOOTING

[Pg 60]

"*Quite tall enough.*"
W.L. Colls, Ph. Sc.

[Pg 61]

CHAPTER IV

SHOOTING

The chief difficulty confronting a host who desires to give his guests good sport lies in the fact that it is no easy matter to get young hand-reared wild ducks to fly well, and I propose in this chapter to endeavour to show how it can best be done.

I say *young* birds, as I think it will be admitted that wild duck, if shot in late October or November, will nearly always fly well. Many sportsmen will, however, for various reasons, not want to keep their birds so long, either on the score of expense or for fear of their straying from home. Young wild ducks hatched about the second week in April should, if properly fed, be in good plumage and fit to shoot by the first week in September; and why, their owner naturally asks, should they go on eating their heads off when they are ready to be [Pg 62] shot and eaten themselves. Partridge driving has not begun and the first edge has been taken off the grouse, so why should not the ducks be shot now; moreover, if fed well they will fetch a good price in the market at this time, as they will be in the nature of a treat so early in the season. The methods of shooting hand-reared wild ducks may be divided into four:—

1. Posting the guns at different spots on the margin of a lake or near it, and flushing the ducks by means of dogs and beaters.

2. Teaching the ducks to take a particular line of flight by means of a horn, and then, without using the horn on the day of the shoot, intercepting them during their flight.

3. Catching the ducks beforehand, liberating them in convenient numbers, and then driving them over the guns.

4. Flight shooting.

To deal first of all with No. 1 method. Let it be imagined that the host is fortunate enough to possess a lake or piece of marshy ground of considerable extent, and bordered by reeds or flags, which form good cover. [Pg 63] Possibly the lake may narrow at some part, and if so our host's dispositions are easy; he places his guns on either shore at the "neck," and if there is room he fastens a punt in the water, midway between the guns on land. A second line of guns might, of course, be placed farther back.

If the lake winds about a good deal the ducks will probably cut across country, and in any case can be easily made to do so by being flagged in or by being fed in a certain direction; there will very likely be some belt of trees in their line of flight, and if so some delightful sport may be had at high birds, the guns being placed in the open and well back from the trees, unless the birds are very shy.

When the dispositions of the host are made, spaniels and keepers will beat the rushes on either side of the lake, driving the ducks over the guns, and the dogs can then be taken to the farther end and a return drive given when the ducks have been over the guns once; the latter will probably have to conceal themselves for this return drive, as the birds will now be more wary, and [Pg 64] many that have not settled at the farther end of the lake may be circling high overhead.

After a time it will probably be necessary to rest the birds for an hour or two for fear of driving them clean away. Don't forget when the shoot is over to have a thorough hunt for dead birds and cripples; the "pick up" is always a big one, as very few birds are missed entirely. The best time to shoot at a high-flying duck is just after he has passed overhead, as then the shot gets behind the feathers and penetrates more easily.

The best shot to use is, I think, No. 4.

The disadvantages of the above plan are: (*a*) all the birds are frightened badly, and some are sure to be lost; (*b*) some birds, which strictly speaking are barely ready, are certain to be shot.

AN INEFFECTIVE CRIPPLE STOPPER

[Pg 65]

Many a good day's sport have I enjoyed with the ducks in India. In the North-West Provinces, where I was once quartered, there are a number of "jheels" or huge lakes, and during the cold weather these are tenanted by countless wild-fowl of nearly every variety. The plan usually adopted is to post the guns some distance apart and where they can best command the favourite feeding grounds of the birds; natives are then sent to different parts of the lake to stir the fowl and afterwards to keep them on the move, should they settle at a distance from where the guns are placed. Well I remember the keen pleasure, not unmixed with anxiety, with which I received an invitation to shoot a celebrated "jheel" which had not been disturbed that season. Ten guns, I was told, were coming. Now I knew that there were not more than half-a-dozen really safe guns in the immediate neighbourhood, and I determined that in my case discretion should be the better part of valour. I accepted the invitation with certain mental reservations.

Arrived at the rendezvous, I found an old friend and good shot; in addition several good fellows, some of whom, though charming from a social point of view, plainly showed by the rather defiant manner in which they handled their guns that they were best avoided on the present occasion. Fortunately for my friend and myself we were rather short of boats, so with apparent good [Pg 66] nature we insisted on staying on shore, where we could get well out of range if necessary. We speedily secreted ourselves amongst some tall reeds, and well away from the direction towards which the fleet of boats was making. One of these, strongly resembling a three-decker, had three guns on board, all of whom stood upright throughout the action. Her we christened the *Man of War*. The smaller craft skirmished in her vicinity, and for two hours the battle raged furiously. No distance was too great, no waterfowl too small or insignificant for their attention; but endurance has its limits, and at last we noticed that even the *Man of War* was silenced, having fired upwards of 600 rounds. Slowly and solemnly the "Fleet" worked its way back to tiffin.

BEFORE THE EVENING MEAL

[Pg 67]

In the meantime my friend and I had some capital sport, killing several pintail before these birds, always the first to leave, had finally departed. In addition we got some grey duck, gadwall, and a number of garganey and pochard. Later, when the boats had all left the "jheel," the fowl slowly began to return, and we now realised with satisfaction that we were well placed. Never have I had better sport or enjoyed myself more, and when at length we were peremptorily informed that the return train was shortly due (and even Indian trains don't wait for one more than half-an-hour), we staggered into the little wayside station, followed by our coolies, carrying enough ducks to feed the station for a week. The second method has now to be dealt with.

Nothing is easier than to accustom the ducks to come to feed at stated times.

At first a horn may be used and then gradually dropped, and in a very short time the birds will know the time of day as well as their feeder does; the latter must be stern with them, absolutely declining to feed them except at the regular hours, one of which will be timed to suit the hour it is intended to commence the shoot. Before commencing this tuition the host will have to select the place from which he wishes the birds to fly, and also the feeding ground which is the end of their journey.

Ducks prefer to rest during the day, and are very fond of shade; provide them, therefore, [Pg 68] if possible, with a plantation on some sloping ground fairly near water, where they can get shelter from sun and wind. I have found willows excellent for this purpose, as by topping they can always be kept at the required height. Such a spot will do admirably as jumping-off place, and here the birds may regularly be expected to rest after their night's wandering in search of food. The next step is to select the feeding ground, which should be some little distance from the spot described; preferably it should be on high ground, so that the ducks in their flight have to pass over some sort of valley situated between the two places. In this valley the guns are placed shortly before the feeding hour, and as that time approaches small detachments of ducks will wing their way across the valley for their meal, and give most sporting shots. It is, of course, essential that the resting-place by day and the feeding ground are not too close together, as if this is the case many birds hearing the firing close at hand may be scared from coming to their food.

A RIGHT AND LEFT

[Pg 69]

After this the birds may be driven back the reverse way, though naturally this practice cannot be repeated more than once or twice in the year, or the birds will be scared away from the feeding ground altogether.

If the host has a piece of water at right angles to the flight of the birds many will scatter after passing the guns and settle; and later on these birds can be driven up and down the water as described in the first method.

One great advantage of the first stage of this plan is that the birds mostly fall on dry land and are easily retrieved. If the ground does not lend itself favourably for high birds the difficulty can be largely overcome by planting a belt of trees and then placing the guns in the open a little distance back; birds inclined to break out at the sides can easily be flagged in.

It is a good plan to run some wire along the slope of the ducks' resting place, as it facilitates their rising at once, and they get into the habit of flying the whole distance instead of walking part of it.

The third system has now to be considered. It is the most artificial of all, and is most suitable in cases where the ground does not [Pg

70] lend itself well for high birds, or the host is not a man of unlimited means, but is fortunate enough to have the shooting rights over a fine stretch of water. The ducks probably vary considerably in size and age, as the owner, not having a large breeding stock, has not been able to put down a large number of eggs at once.

The time has, however, come, when he has sufficient to give his friends a very nice shoot. It is, of course, undesirable to frighten or damage either the pinioned or immature birds, and these latter will have to be sorted from those which are fit to kill.

The first step will be to accustom the birds to feed inside a wire enclosure, with some dark building, such as a barn or stable, at one end of the enclosure, and connected with it by means of a door. The birds all having been coaxed inside the enclosure to feed, shut the door of the enclosure quietly, and gradually drive the birds into the dark building. Here the birds will be left all night, and owing to the darkness will not damage themselves. A certain amount of ventilation and some water will be necessary.

AT THE END OF THE DAY

[Pg 71]

It is a bad plan to give them any food beyond a light meal the evening they are caught, and certainly nothing next morning, as otherwise they will fly badly and heavily when liberated.

Next morning, those ducks that are fit to shoot will be separated from the pinioned birds and those that are immature, and these latter can be conveyed in hampers to any convenient building, and fed.

They will be kept in confinement during the shoot.

Now for the shoot itself. The man who feeds the ducks has for a considerable time trained the ducks to fly in and out of the paddock or yard, in which the enclosure is situated. This is easily done by stretching a piece of wire, which can be gradually increased in height, across the boundary of the paddock into which the ducks come for their food. They soon get accustomed to this wire, and realise they will get no supper if they don't take the trouble to fly.

As has already been mentioned, the owner of the ducks has the shooting rights over a fine piece of water, and on this water, and in [Pg 72] the cover which grows round it, the birds pass the time between their feeding hours. There is sure to be a line of willow trees of some sort or other near the water's edge, and it is over these the ducks must be made to fly. Provided that a small clump of low willows, or other cover, is planted some distance from the rearing field, with the high willow trees standing between the two, it is quite easy, by occasionally feeding in this little cover, to accustom the birds to look on it as their sanctuary, and when liberated from their enforced confinement they will make straight for it, and over the tops of the intervening trees. All that has to be done now is to place the guns between the tall willow trees and the little cover, but well in the open, so that the ducks may see them and be induced to rise higher in consequence.

A little false cover can now be put along the wire before alluded to at the edge of the rearing field, to make the birds rise better, and to prevent the guns from getting any inkling of your plan of operations.

COMING ON A SIDE WIND

[Pg 73]

All is now ready, and at a given signal the birds which have been shut up all night will be liberated in detachments of varied numbers, first from the dark building, and secondly from the wire enclosure. Thoroughly frightened with their unaccustomed imprisonment, they take wing at once, and make the best of their way to the sanctuary, giving the guns most sporting shots. Should the wind be across their line of flight to the sanctuary, you will of course have to flag them in, as ducks always rise up wind, and love to fly against it; nothing they detest so much as getting their feathers ruffled. It will be found that they always fly best on a dull stormy day.

The piece of water behind the guns should preferably run at right angles to the line of flight of the birds from the paddock to their sanctuary, as after the birds have passed the guns they will split up right and left, and settle at one end or the other. The guns will next be placed so as to command the water from bank to bank, one of them being placed, if necessary, in a boat moored for the purpose in midstream.

The ducks are now driven over the guns again, down wind for choice, and this can be [Pg 74] followed by a return drive, which ends the day's sport.

An hour later some one must search the lake thoroughly for cripples, and when this has been done the breeding stock and immature birds should be released.

A modification of this plan may be tried, though I do not recommend it. Instead of the birds being liberated from the enclosure as already described, they are caught, placed in hampers, and conveyed to some convenient spot at a distance from home, and then liberated in the numbers required. The birds naturally fly straight home, and sometimes fly well. Care must be taken to set them free amongst surroundings they know, otherwise they are cowed like a rabbit liberated away from its burrow. It is also advisable to place some obstacle across their line of flight, and about sixty yards in front of the guns, so as to make the birds rise well. The last plan has the obvious disadvantage that the ducks must be cramped to a certain extent by their imprisonment in the hampers, and it savours too much of the artificial to ever prove a complete success. On the other hand, the method described as the third works well; the birds are not crowded, but on being liberated are glad to escape; they are frightened and mean to fly well: but best of all your breeding stock and immature birds will, if this principle be adopted, know nothing of the shoot, and on being let go, will settle down in a very few hours and will assist in taming those birds which have been shot at but escaped. Whatever you do be careful to conceal all your plans from your guns, when artificial methods are adopted; the day is always more enjoyable if the guests cannot see how their host manages matters.

LADIES IN WAITING

[Pg 75]

Ducks are extraordinarily good barometers, and by their behaviour on the water invariably give warning of coming rain or storm.

No one who has kept wild ducks long has failed to notice their peculiar uneasiness before bad weather.

Suddenly one bird with outstretched wings will dash madly on the surface of the water, and behaving much in the same way as a flapper chased by a dog, throw itself into the air, and dive suddenly on alighting again: in a moment this is taken up by every bird on [Pg 76] the water, until one sees the extraordinary sight of two or three hundred ducks behaving just as if they were mad. They dash in all directions and appear quite unable to control themselves. When all this is noticed there is pretty sure to be rain within twelve hours.

The last but by no means least sporting form of duck shooting must now receive a little attention. I allude to Flight Shooting. As winter comes on the ducks' natural instincts have begun to assert themselves, and regularly at dusk, heads will go up, and a peculiar uneasiness manifest itself: very shortly the birds will fly off, after one or two preliminary circles round, to the feeding ground they have selected, though if properly fed they will not go far. All that has to be done is to observe where the ducks feed, and place the guns either in the line of flight between the birds' home and their feeding ground or round the feeding ground itself.

No sport is more fascinating than this—the absolute solitude, the dull red glow of the light fading in the west, gradually getting fainter and fainter, the light shiver of the reeds, as a breath of wind rustles through [Pg 77] them, and best of all the whistle of beating pinions high overhead, betokening the welcome intelligence that birds are circling round, and making a full inspection of the feeding ground before alighting. Don't move now whatever you do, your retriever, sitting close at your side, will move his head quite enough, without your stirring as well: if you watch him you will soon get a pretty good idea as to where the birds are.

Presently the noise becomes louder, and then with a loud "swish" the birds come right at you. Throw up your gun quietly and quickly and fire at once—don't dwell on your aim, and let us hope that the dog has no difficulty in retrieving a bird that was evidently cleanly killed.

Ducks, like other birds, always alight facing the wind, and this fact must be borne in mind when selecting the stand. Should there be no wind to speak of, it is best to face the fading light, unless the ducks are known to make a practice of coming from one particular direction.

They are most capricious birds, here to-day, and gone to-morrow, but this all adds to the [Pg 78] fascination of the sport. I remember once killing eight ducks at a particular spot one evening, and not even getting a shot the next, although there were hundreds of ducks in the neighbourhood. Very different sport to this does one get in the East. The man who goes Flight Shooting there is almost certain of good sport, provided he knows what he is about. Well I remember a certain evening in Upper Burmah. It was at Alon on the river Chindwin, and during the last Burmese war.

We were not supposed to go far from the Fort, but if we took an armed escort with us, no objection was raised.

There was a large "jheel" about two miles from the Fort, which was much overshot by the small garrison quartered there, and during the day little could be seen on its surface besides a few whistling teal, a duck that gives poor sport, and is only just worth eating.

I discovered, however, that at dusk hundreds of ducks returned to the "jheel" from all directions, remaining there probably until dawn.

[Pg 79]

Followed by my soldier servant as an escort, I made my way to the "jheel," and having made our passage in one of the frail boats, known as "dug-outs," we eventually arrived at a small island which I had selected for my stand.

Never shall I forget that evening. For about twenty minutes I shot nearly as fast as I could load, and not too well, I am afraid.

Ducks of several different varieties were coming fast, and at all angles and elevations.

Many an old sportsman will understand my difficulties. I had of course no second gun, no ejector, and at times I utterly forgot the motto "Festina lente."

At last it was over, and I went home moderately satisfied with about five-and-twenty ducks, leaving, alas! a large number unpicked, as we had no dog.

When your shoot is over, and the season is drawing to its close, the only work left amongst the ducks is to select the breeding stock for next season.

The best to keep are long and well-furnished birds, as they always fly better, and lay more eggs than the short thick-set variety: [Pg 80] they should have rakish-looking heads, with long bills, chrome yellow tinged with green in the case of the drake, and dull brown fringed with bright orange in the case of the duck. The eyes should be set high in the head, and the head itself *appear* to be slightly angular in appearance, and not too round at the crown. I believe in fairly light coloured birds, as I have always noticed that any strange birds that arrive appear lighter in colour than my own, and I think that the darker and coarser birds do not fly so well. In any case get rid of all short thick-set birds—they will do for the table, but not for sport.

In taking leave of my readers, I hope that I may have been fortunate enough to secure a little of their interest, and that this book may prove of some assistance to those who, like myself, love wild duck, and consider a few hours spent daily in their company an education and a treat.

www.ingramcontent.com/pod-product-compliance
Lightning Source LLC
Chambersburg PA
CBHW030511220526
45464CB00006B/2750